HISTORY OF INVENTIONS

Music Technology

From Gramophones to Music Streaming

Tracey Kelly

BROWN BEAR BOOKS

Published by Brown Bear Books Ltd
4877 N. Circulo Bujia
Tucson, AZ 85718
USA

and

Unit 1/D, Leroy House
436 Essex Rd
London N1 3QP
UK

© 2019 Brown Bear Books Ltd

ISBN 978-1-78121-454-1 (library bound)
ISBN 978-1-78121-470-1 (paperback)

All rights reserved. No part of this book may be reproduced, stored in a retrieval system, or transmitted, in any form or by any means, electronic, mechanical, photocopying, recording, or otherwise, without the prior written permission of the copyright holder.

Library of Congress Cataloging-in-Publication Data available on request

Text: Tracey Kelly
Designer: Supriya Sahai
Design Manager: Keith Davis
Picture Manager: Sophie Mortimer
Editorial Director: Lindsey Lowe
Children's Publisher: Anne O'Daly

Picture Credits
Front Cover: iStock: Dragon Images.
Alamy: Pat Canova, 10, ClassicStock, 8; Brown Bear Books: 4; iStock: Fertnig, 12, ikushakof, 13c, jakkapan21, 14, jgroup, 11, Philartphace, 6br, Skrow, 19, trekandshoot, 16, Viktorus, 13b; MTV: 16; Public Domain: Gatech Education, 20, Meister des Codex (Grundstockmaler), 5l, NPM.gov.tw/British Museum, 5tr, Nuberger, 6t, Stwity, 18; Shutterstock: Steve Collender, 15, Glepi, 9, Sergej Razvodovskij, 17, Anatolii Riepin, 21, velora, 7.
Key: t=top, b=bottom, c=center, l=left, r=right

Brown Bear Books have made every attempt to contact the copyright holder. Please email licensing@brownbearbooks.co.uk if you have any information.

Brown Bear Books has made every attempt to contact the copyright holder.
If you have any information, please contact: licensing@brownbearbooks.co.uk

Manufactured in the United States of America
CPSIA compliance information: Batch#AG/5624

Websites
The website addresses in this book were valid at the time of going to press. However, it is possible that contents or addresses may change following publication of this book. No responsibility for any such changes can be accepted by the author or the publisher. Readers should be supervised when they access the Internet.

Contents

Sound of Music 4
Gramophone 6
Radio 8
Record Player 10
Hi-Fi 12
Cassette Tape and Deck 14
Compact Disc 16
Digital Music 18
Future Music 20
Timeline 22
Glossary 23
Further Resources 24
Index 24

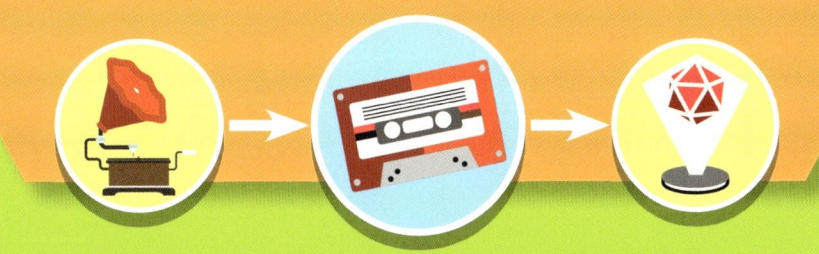

Sound of Music

People have always made music. But they didn't always have machines to record it! People went to concerts or the theater. They played instruments and sang at home.

Amphitheaters
The ancient Greeks built amphitheaters. These were big outdoor theaters. Amphitheaters were built in a special way. They made the singers and musicians sound louder.

Ancient Instrument

A lyre is an ancient instrument. It is thousands of years old. You strummed or plucked the strings. People still play lyres today.

Music on the Move

Minstrels were traveling musicians. They moved from place to place. Minstrels sang songs and played music. They shared news from far away.

Read on ... to find out about how people recorded music to play again and again.

Shellac Records

Shellac is a sticky material. It is made by insects. It starts off being soft. Shellac was flattened into a disc. After a groove was cut, the shellac hardened.

HORN

Gramophone
You turned the handle. That made the record go around. A needle ran along the groove. Music played through the big horn.

HANDLE

NEEDLE

RECORD

Radio

Radio waves send sounds through the air. They are invisible waves, like heat and light. Radios pick up the radio waves. They turn them into sounds. You tune the radio to listen to different programs.

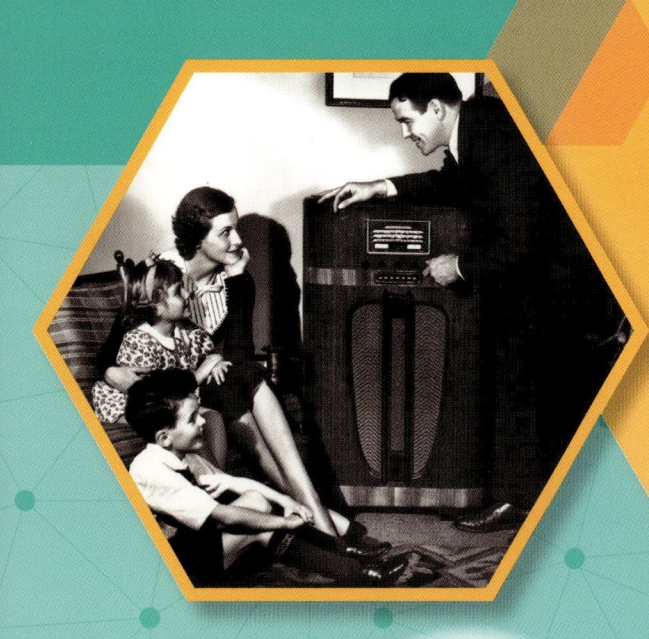

Kids' Radio
Children listened to shows on early radios. The shows were stories like *Little Orphan Annie*. The whole family listened together!

- RADIO
- LOUDSPEAKER
- STATIONS
- DIAL
- VOLUME
- ON/OFF SWITCH

The first radios were very big.
They were made of wood.
You turned the dial to find a show.

Record Player

The record player was an electric gramophone. It played records. An electric motor turned the record around. You didn't need to turn a handle, like on a gramophone.

Jukeboxes
Restaurants had jukeboxes. This was a kind of record player. You could play your favorite song. You put money in. Then, the record dropped down and played.

RECORD

NEEDLE ARM

KNOB

SPEAKER

New records were made from vinyl, a type of plastic. A needle arm lifted the needle. It went on and off the record to play it.

Hi-Fi

A hi-fi was a music system. Hi-fi is short for "high fidelity." Fidelity mean "faithful." Music sounded true to life. It was just like being at a concert!

Stereo Sound
Most hi-fi systems had two speakers. The music was split into two parts. Different sounds came from each speaker. This was called stereo sound.

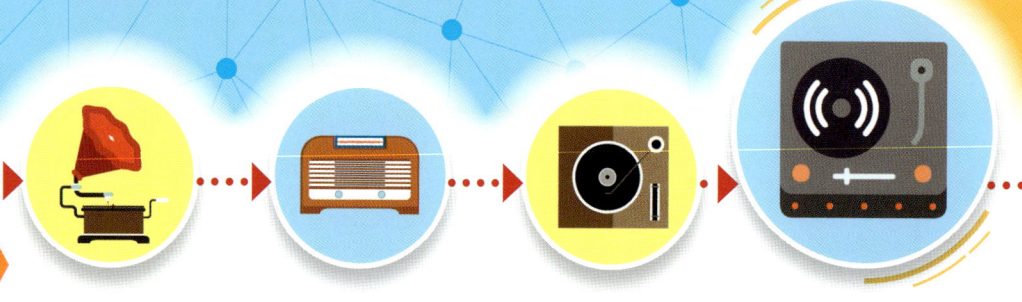

A hi-fi had different parts. It had a turntable. This played vinyl records. It had a radio. An amplifier made the music louder. Music played through speakers.

SPEAKER

TURNTABLE

RADIO

13

Cassette Tape and Deck

Cassette tapes stored music on a long, thin tape. The tape was rolled up inside a small case. Cassette tapes were easy to carry around. You played them on a cassette deck.

Cassette Deck
Cassette decks were smaller than record players. You could play music on them. You could record your own music, too.

Boom Box

A boom box had two cassette players. It often had a radio, too. It was heavy. But you could carry it around. You could listen to music anywhere!

RADIO

SPEAKER

CASSETTE

BUTTONS

15

Compact Disc

A compact disc is also called a CD.
The disc was made of plastic and metal.
One side was shiny. One side was dull.
The music was stored on the shiny side.
A laser beam made the music play.

Music TV
MTV was a TV station. It played only music videos. Singers and bands made videos to go with their songs. MTV became very popular!

COVER

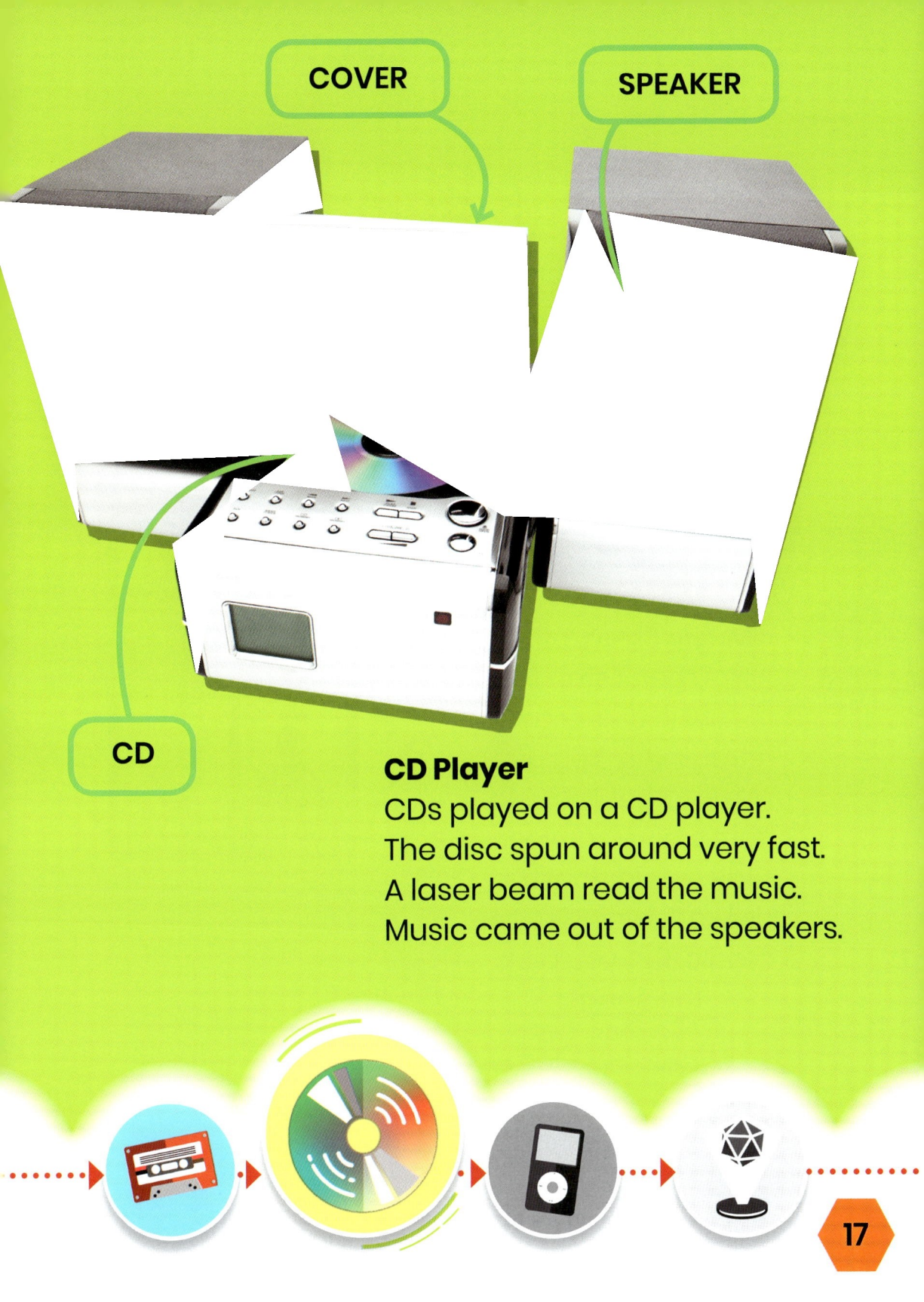

SPEAKER

CD

CD Player
CDs played on a CD player.
The disc spun around very fast.
A laser beam read the music.
Music came out of the speakers.

Digital Music

An MP3 was a computer file. It stored audio. Audio is another word for sound. Each song was on a different MP3. You could store lots of songs in a small space. MP3s played on an MP3 player.

Music Streaming

You can listen to music online by streaming. You use Wi-Fi to get it. Music streaming services play music nonstop. It is like listening to the radio. But you pick the songs!

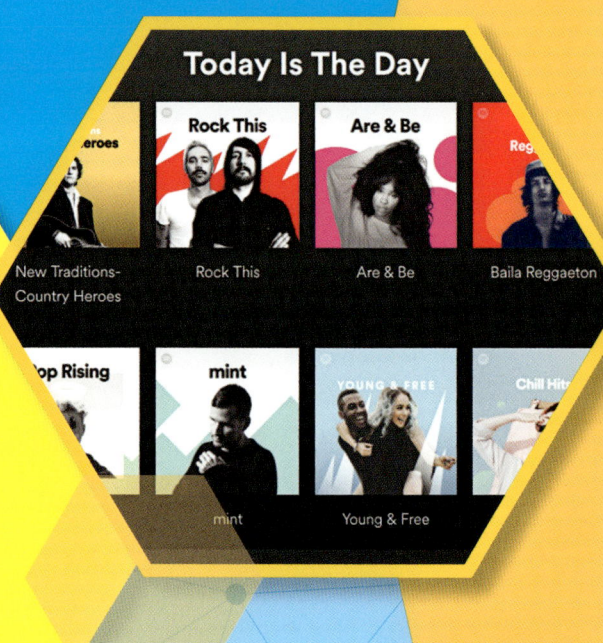

SEARCH

SONGS

MP3 Player
An MP3 player stored thousands of songs. You bought MP3s from an online store. Then, you downloaded them to your MP3 player. You listened through headphones.

CONTROLS

HEADPHONES

19

Future Music

Music may be different in the future. Artists will record for new formats. You will be able to listen in new ways. What do you think will happen next with music?

Robot Music
Shimon makes his own music. He is a robot. He learned thousands of songs. He can write his own tunes, too!

VR Music

Some apps will let you play VR music. VR means virtual reality. This is a pretend world made up by a computer. You put on a headset. You see videos and hear music. You can sing along "on stage" with your favorite stars!

HEADSET

HEADPHONES

Timeline

1877 — The phonograph is invented by Thomas Edison. It is the first sound recorder.

1887 — The gramophone is invented by Emile Berliner.

1895 — Guglielmo Marconi develops wireless telegraphy.

1930s — Most homes have a radio.

1931 — Stereo sound is invented.

1941 — FM radio begins in the United States.

1950 — Electric record players become popular.

1963 — The compact cassette tape is invented.

1960s — Hi-fis are used in homes.

1981 — The MTV channel launches.

1982 — The first compact disc (CD) player is launched.

1997 — An MP3 player is launched.

2001 — The first online music store launches.

early 2000s — Music streaming services become popular.

2017 — Shimon the robot makes music at Georgia Tech, Atlanta.

Glossary

amplifier A device that makes sound louder.

audio Another word for recorded sound.

CD (compact disc) A flat disc that stores music.

digital Information that is stored as the numbers 0 and 1. Computers use digital information.

fidelity A word for faithfulness.

format The way a sound recording is made to be played.

groove A spiral cut in a record disc.

laser beam A powerful beam of light.

lyre A stringed instrument that is like a harp.

MP3 player A music player that plays music files.

radio waves Electronic waves used for sending messages and sound over long distances.

shellac A sticky liquid that comes from insects. It was dried and used to make records.

streaming Playing music or video as it is sent over the Internet.

virtual reality A make-believe world made up by a computer.

Wi-Fi A wireless connection to the Internet.

Further Resources

Books

Thomas Edison Barbara Kramer, National Geographic Children's Books, 2015

Making Contact!: Marconi Goes Wireless (Great Idea Series) Monica Kulling, Tundra Books, 2016

Websites

Find out more about sound recording: wiki.kidzsearch.com/wiki/Sound_recording

Listen to the early sound recordings made by inventor Thomas Edison: nps.gov/edis/learn/photosmultimedia/the-recording-archives.htm

Index

amphitheaters 4
amplifier 13
cassette tape 22
CD 16, 17, 22
compact disc 16, 22
data 24
gramophone 6, 10, 22
hi-fi 12, 13
lyre 5

minstrels 5
MP3 18, 19, 22
MP3 player 18, 19, 22
MTV 16, 22
streaming 18, 22
radio 8, 9, 13, 15, 18, 22
record player 10
VR 21
Wi-Fi 18